WHAT IS OPHIOLOGY?

Ophiology is the scientific study of snakes.

The scientists who study snakes are called **OPHIOLOGISTS**.

Words that are tricky to understand are in **bold**. Find out what they mean in the glossary.

Words that are difficult to say are in *italics*. Find out how to say them at the back of the book.

CAN SNAKES PREDICT EARTHQUAKES?

DISCOVER THE SCIENCE BEHIND OPHIOLOGY
(oh-FEE-oh-luh-jee)

Written by Eliza Jeffery
Illustrated by Denis Alonso

Over time, snakes have been pretty good at surviving, no matter where they live. They have cleverly learned how to **adapt** to changes in their **habitat**, and to **predators** that might **attack them!**

The inland taipan has the most toxic **venom** of any snake in the world. One drop can harm over 100 humans! Scientists called *ophiologists* discovered this snake's venom has changed over time to be most deadly to **warm-blooded** animals.

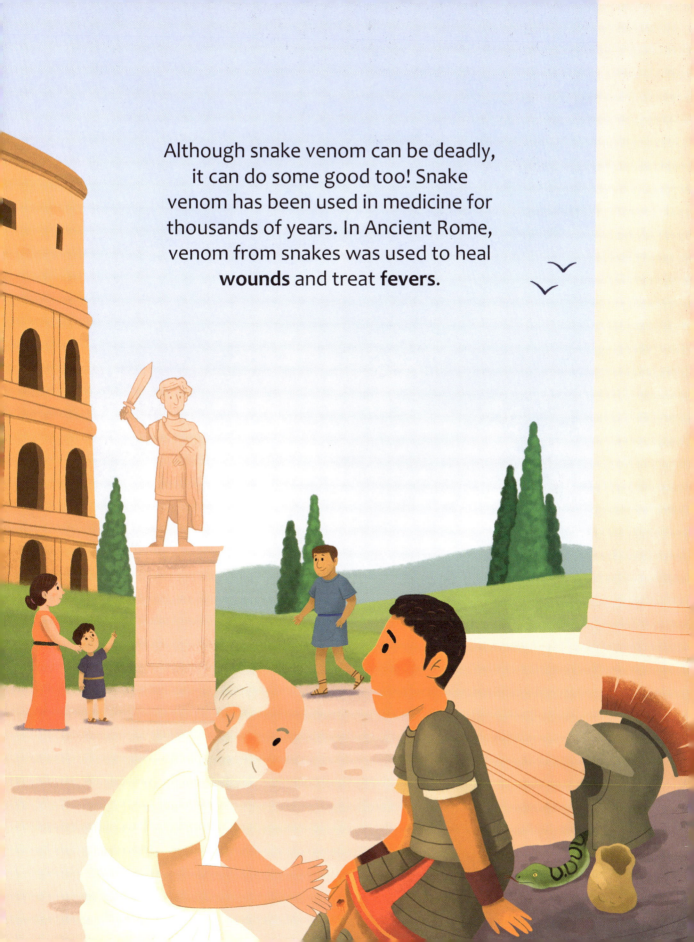

Although snake venom can be deadly, it can do some good too! Snake venom has been used in medicine for thousands of years. In Ancient Rome, venom from snakes was used to heal **wounds** and treat **fevers**.

Snake venom is used in our medicine now too!
Not only can it heal wounds, it can also cure
diseases and treat all sorts of pain.

Scientists are continuing to discover new ways of
using venom to create cures. It has a really
valuable role to play in medicine today.

Snakes don't just live on land... they can even **live in the sea!**

Ophiologists discovered that yellow-bellied sea snakes can take in most of the **oxygen** they need to live through their skin! They have also adapted to create an "S" shape with their bodies when they swim – this helps them move through water easily.

There are even **snakes that can fly!**

The paradise flying snake jumps from extreme heights and glides through the air, moving and flattening its body to help it land safely. Ophiologists believe this is an example of snakes using their **environment** to escape predators, and hunt **prey** too!

A snake's connection to their environment also helps them hunt for prey. Snakes have a special **infrared system** that uses heat to track down prey, which helps them catch animals that think they're hidden from sight!

Snakes have nostrils that allow them to sniff out prey, but they have also adapted to **smell using their tongue!**

Once they have their prey, it's time for snakes to eat. Snakes, like the green anaconda, have incredibly flexible jaws – they can eat prey much larger than their head!

They also have many small teeth that curve backward to help them hold onto prey. As impressive as this is, snakes must be careful – they have been known to **explode if they eat too much food!**

As snakes grow, they have the ability to **shed their skin!**

Although other animals can shed their skin, snakes are one of the only ones that shed all at once, leaving snake-shaped socks behind.

Snakes use their environment to hide from predators. The Sahara sand viper has scales that allow it to perfectly blend in with its sandy surroundings...

the timber rattlesnake has green and brown scales to blend in with the forest floor...

and the head of the Madagascar leaf-nosed snake looks like, you guessed it, a leaf! Snakes are **experts at camouflage!**

Perhaps the most perfect habitat for snakes to hide out in is the constant warmth of a **tropical** rainforest. The Amazon Rainforest has over 300 **species** of snakes living there.

No matter how different these snakes may look, they have all adapted to thrive in this impressive environment.

Brazilian blind snake

False water cobra

Scientists are always discovering amazing new things about snakes.

In 2006, ophiologists in China began to notice snakes were acting in a curious way – they were headbutting and throwing themselves at walls!

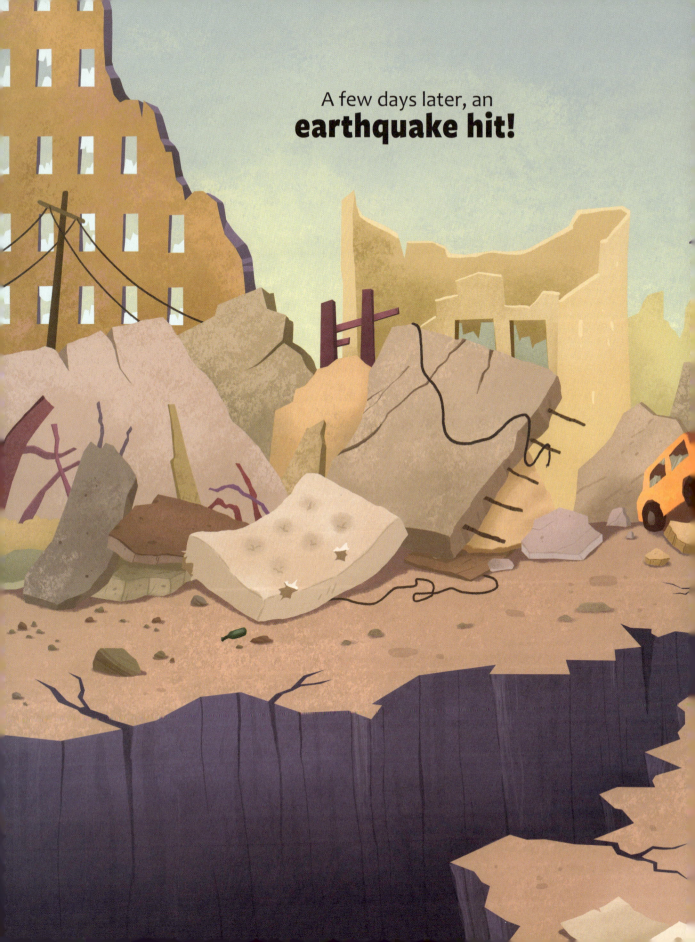

Ophiologists found that these snakes were reacting to the earthquake that was on its way! They did this by sensing the smallest of ground movements in their body. Of all the animals on Earth, snakes seem to be the most sensitive to earthquakes.

Because of this event, ophiologists began to watch snakes in earthquake-prone areas and track any unusual habits. This was an amazing discovery that proved just how impressive snakes really are. Alongside all their other amazing abilities, snakes can predict earthquakes!

Record-breaking SNAKES

Snakes are leading in their ability to connect with their surroundings. What other ways are snakes setting records, and standing out from the rest?

The heaviest snake is... THE GREEN ANACONDA!

Not only are green anacondas the largest snakes, they're the heaviest too! The heaviest ever recorded weighed 550 pounds (250 kg), which is around the same weight as a grand piano!

The smallest snake is... THE BARBADOS THREADSNAKE!

The record for smallest snake goes to the Barbados threadsnake, which only grows to be around 4 inches (10 cm) long! These tiny snakes are as thin as a piece of string.

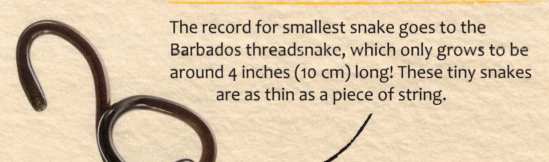

The longest snake is... **THE RETICULATED PYTHON!**

The longest snake species alive is the reticulated python, which can grow up to 32 feet (10 m) long. That means this snake can be around the same length as a school bus!

The friendliest snake is... **THE CORN SNAKE!**

Corn snakes are known for being gentle and calm, making them a perfect species to keep as pets! They are small and don't tend to bite.

The fastest snake is... **THE BLACK MAMBA!**

As if snakes aren't already scary when they slither, there is one species that can move extremely fast! The black mamba has been recorded as moving at approximately 12 mph (19 km/h).

Slithering
SNAKE FACTS

Snakes may be able to predict earthquakes, but ophiologists have discovered lots of other fascinating things about them too…

SNAKE-ONLY ISLAND
Off the coast of Brazil lies an island called *Ilha da Queimada Grande* – also known as Snake Island – which is full to the brim with snakes! Nobody is allowed to visit, apart from ophiologists who carry out research there.

A FEAR OF SNAKES
Ophidiophobia is the fear of snakes, and many scientists believe this fear comes from our age-old understanding that snake venom can be deadly. Research suggests that 1 in 3 people have a phobia of snakes.

LACKING EYELIDS

Snakes don't have eyelids like we do, so they never blink! They sleep with their eyes open, making it tricky to tell if snakes are watching you or not...

SNAKES LIKE TO SUNBATHE

Snakes will often sit in the sun at the start of each day to keep their body temperature high. Sunshine helps to keep snakes' bones strong and healthy.

VENOMOUS VS NON-VENOMOUS

Ophiologists have discovered that most snakes with round **pupils** are not **venomous**, and most snakes with slitted pupils are venomous. However, there are exceptions!

GLOSSARY

Adapt – when a living thing develops special features or skills to help it survive in its environment.

Camouflage – the way animals blend in with their surroundings so they can't be seen easily.

Diseases – conditions that cause part of a living thing to no longer work properly.

Environment – everything that is around us.

Fevers – a condition where your body gets very hot because you are sick.

Habitat – the places where animals or plants live.

Infrared system – a way to see heat that animals give off.

Oxygen – an invisible gas in the air that plants produce, and people and animals need to breathe.

Predators – animals that hunt other animals for food.

Prey – an animal that is hunted by other animals for food.

Pupils – the black part in the middle of the eye.

Species – a group of living things that share characteristics and features. For example, a cobra and an anaconda are different species.

Tropical – places that are very warm and often rainy, such as rainforests or beaches.

Venom – a poison some animals use to hurt or kill their prey (see left).

Venomous – a creature that can produce venom (see above).

Warm-blooded – animals, such as humans and birds, that keep their bodies warm even when it's cold outside.

Wounds - injuries to the skin.

HOW DO I SAY?

Ilha da Queimada Grande
EEL-yah dah
kay-MAH-dah GRAN-dey

Ophidiophobia
OH-fid-EYE-oh-FOH-bee-ah

Ophiologists
oh-FEE-oh-luh-jists

Ophiology
oh-FEE-oh-luh-jee

THE BIG QUESTIONS ANSWERED

This is more than just a series of books; it is a complete resource.
Accompanying each book is a variety of FREE material to engage curious kids with science.

www.thebigquestionsanswered.com

Use the QR code to visit the website, download free resources, and discover other books in the series.

On the website, find out incredible things about ophiologists, including what they do, some of their greatest discoveries, and the people who have made a difference in this field of science.

The material is also available for home or classroom use, supporting all the information in this book.

Teachers' & Parents' Resources
With discussion prompts and questions, extra information, and facts around key topics.

Young Ophiologists' Activity Pack
Fun activities for wannabe snake experts, including creative writing, drawing, word searches, and much, much more.

BEETLE BOOKS

The Big Questions Answered is published by Beetle Books.
Beetle Books is an imprint of Hungry Tomato Ltd.

First published in 2024 by Hungry Tomato Ltd
F15, Old Bakery Studios, Blewetts Wharf, Malpas Road,
Truro, Cornwall, TR1 1QH, UK.

ISBN 9781835691373

Copyright © 2024 Hungry Tomato Ltd

No part of this publication may be reproduced, stored in a retrieval system, or transmitted in any form or by any means, electronic, mechanical, photocopying, recording, or otherwise, without prior written permission of the copyright owner.

A CIP catalog record for this book is available from the British Library.

With thanks to:
Editor: Millie Burdett
Editor: Holly Thornton
Senior Designer: Amy Harvey
The team at Beehive Illustration

Printed and bound in China.

Picture Credits:
(t = top, b = bottom, m = middle, l = left, r = right)
Shutterstock: kamnuan 33br; kaveetha kumudumalee 34 mr; Muhammadsaifalkhan 35bl; Nynke van Holten 33t; SeventyFour 34bl; somyot pattana 32bl; Stephanie426 35mr; reptiles4all 35tl.